T0199206

GETTING THIS FAR

Between a Planck and a Parsec

Peter Cockrum

To order additional copies of this book, contact:
Xlibris
1-800-455-039
www.xlibris.com.au
Orders@Xlibris.com.au

ISBN: Softcover 978-1-9845-0608-5
Hardcover 978-1-9845-0609-2
EBook 978-1-9845-0607-8

Library of Congress Control Number: 2020908369

Print information available on the last page

Rev. date: 05/22/2020

GETTING THIS FAR

Between a Planck and a Parsec

INTRODUCTION

Who am I and why am I writing this?

Well, I aim to have this book in print for my 75th Birthday (24 August 2020).

I am proud Father to Elise and Lyndal, Father in Law to Phill and Dennis and very proud Grandfather to (L to R in above photo) Grace, Andre, Oriel and Rory.

I was born at the very end of World War II, in the Angaston Hospital in the Barossa Valley in South Australia. My father (Christened Herbert William Porter Bushel Cockrum and known to Elise and Lyndal as "Gaa") was serving on the staff of General MacArthur in Brisbane and did not want Mum (Clare Eileen Veronica Spurling known as "Nanna") to give birth in Brisbane (the Japanese were a frightening presence to the North at that time and Dad was in the Intelligence service) so my Mother went to stay with her younger sister Celia Harris (they were respectively the seventh and eighth children in the Spurling family) in Nuriootpa, in the Barossa Valley. The neighbouring town of Angaston had the only hospital in the Region.

My father was de-mobbed from the Army a few weeks after I was born and joined my mother in Nuriootpa then moved to Mildura where he and another man jointly ran the Sunraysia Daily Newspaper – Dad as Reporter and Editor and the other man as Printer. After a few years Dad moved the family to Melbourne taking a job at The Age, which he held for the rest of his working life.

We lived first at 56 Regent Street, Oakleigh, then at 14 Wellman Street, Box Hill South and finally moving to 30a Centre Road, Vermont, where my parents lived out the rest of their lives and from where I Married Robyn Pratt.

I remember nothing of my early days in Nuriootpa or Mildura. I have hazy memories of living in Oakleigh and recall moving to Wellman Street where we lived through all my school years. I first went to school in Oakleigh then transferred to Mont Albert Central School (Central Schools don't exist anymore), starting in Grade 3 and moving to Box Hill High School in Form 2, then briefly to Camberwell High School because Box Hill High did not offer Biology at Matriculation level.

I was a very poor student during my school years and, as a result of a teenage tantrum on my part, I left Camberwell High School without finishing my Matriculation and applied for several jobs.

I received two offers – one for a Laboratory Technician at Alice Springs Hospital and the other for a Junior Technical Assistant in the CSIRO Division of Animal Health. I went to a very intimidating interview at the Divisional headquarters in Parkville and was offered the job.

The job was an Assistant position to work with the Division's Analytical Chemist, John Bingly however he was taking a sabbatical in Argentina so his Assistant position was being loaned to Claude Culvenor at the Division of Organic Chemistry, who headed chemical studies into poisonous plants which caused disease in farm animals, collaborating with a group of staff at Division of Animal Health.

The terms of my appointment were that at the end of that year I could choose to move back into Division of Animal Health when John Bingly returned or elect to formally transfer to Division of Organic Chemistry and continue to work with Claude Culvenor. I chose the latter course.

Claude was a role model and surrogate father-figure for me, and I worked with him for more than 20 years until he retired and modelled my work ethic on him. I rose from Junior Technical Assistant to Technical Officer then moved to the Professional ranks and ended up at the career ceiling of Principal Experimental Scientist. To have progressed further I would have had to resign and take a fellowship position whilst obtaining a PhD Degree which held no appeal since my salary would have been less than half what I was then earning and Elise and Lyndal were at Lowther Hall and incurring substantial fees.

I was very naive in those early days and I recall looking for a book on the shelves in Claude's office one day and coming across his thesis for the Degree of Doctor of Philosophy at Oxford University in which he, as author, designated himself C.C.J. Culvenor, PhD. I was puzzled, thinking that you had to have your thesis accepted before you could claim the Degree so I asked my colleague Les Smith and he patiently explained that this was for Claude's second doctorate degree (he already had a PhD from Melbourne University). A doctor of philosophy from Oxford University is the only such degree worldwide for which the abbreviation DPhil. can be used. Thus Claude became CCJCulvenor PhD, DPhil and a few years later, when he was awarded a Doctor of Science from Melbourne University he became one of a handful of CSIRO Scientists with a triple doctorate: C.C.J.Culvenor PhD DPhil DSc. He is currently 93 and I still keep in touch with him although he is now locked down in a Nursing Home due to Covid19.

Having said I was a bad student in my school days, vanity compels me to say I improved with time and when, in the 1980s, RMIT brought a world famous toxicologist from the Karolinska Institute in Copenhagen to Melbourne to introduce a Master's Degree in toxicology to

Australia, I was In the inaugural student intake and topped the course in practical subjects – the oldest student (one year older than me - we were both in our 40s) who worked in the Department of Health came top in the regulatory aspects of toxicology. I reconfigured my thesis into a paper presented at an international symposium[1].

I retired from CSIRO in June of the year I turned 55, the Millennial Year of 2000. Oddly, as a Government employee one's Superannuation pension is calculated differently if you finish up at 54 years and 11 months rather than 55 years and is substantially higher, so I retired on 30th June although I did not turn 55 until August that year.

I immediately enrolled in a course to qualify as a travel agent hoping to see my days out as a local suburban Travel Agent, facilitating my desire to travel. I really enjoyed that course, topping it too, but needed six months experience in an Agency before qualifying for a licence. When I applied for jobs, I had no offers because I had no retail experience. To get some experience I looked for a sales job and found one selling timeshare which offered on the job training. To everyone's surprise I was good at it and was "Salesman of the Year" twice on the Richmond sales deck of Holiday Concepts P/L. They then made me a manager and I was then earning too much money to even consider looking for a job as a Travel Agent, and anyway I had realised with concern that the internet was sending lots of small Agencies broke.

Robyn and I had owned timeshare at a place called The Cumberland, Lorne (not part of the Holiday Concepts Group), so I believed in the concept but had some misgivings about the ethics of the "Hard Sell" timeshare was notorious for, and after a few years we parted company.

I then worked for an Adelaide headquartered roofing company, Modern Roofing, and I am proud to say that after 6 months I was sacked! The ethics and corruption in that company made timeshare look benign and I spoke my mind.

The final episode in my retail career was several mainly pleasant years with a bedding company called "Beds for Backs" from which I finally separated in 2010.

1 Cockrum, P.A., Petterson D.S. and Edgar J.A. (1994) Identification of Novel Phomopsins in Lupin Seed Extracts. In: Plant-associated Toxins: Agricultural, Phytochemical and Ecological Aspects, pp232-237. (Eds .S.M. Colegate and P.R. Dorling). CAB International, Wallingford, England.

Out of all that, I have met many acquaintances but retain only a few really close friends: Peter Bryce from my days at St Peter's, Eastern Hill, Franz Schnellmann (well known to Elise and Lyndal as a near neighbour in Hazelwood Court, St Albans, Brian ("Pip") Cooper my fellow but senior (2years in age but vastly more in Retail Sales experience) and "Chami" Mendez, both Managers at Holiday Concepts and Tibor Dehaney, whom I met and mentored when he joined Beds for Backs. In recent years the "Thursday Crew" at the Fire Services Museum of Victoria have joined that number: Dave, 'Molly', Robbo, Graham and the "Johns", especially "Johnny Have-a-chat" OBE and John Schintler OBE (**O**ver **B**loody **E**ighty).

And that is me in a nutshell.

I was never a good communicator so what follows is a bit of an insight into what occupies my mind.

Chapter I

BETWEEN A PLANCK AND A PARSEC

The metre is the measure of man. I estimate that well over 90% of all *Homo sapiens* ever to have lived had a stature of between one and two metres. Where does that place us in the "scheme of things", that is to say, the universe?

Well, surprise, surprise, we are somewhere in the middle!

This statement embodies what I love about science. Something so simple contains such complexity. To quantify these limits, we need to call on the two great scientific theories of the 20th Century: Relativity and Quantum Theory.

Relativity sprang almost fully formed from the brain of Albert Einstein a little over 100 years ago in two publications: Special Relativity in 1906 and General relativity in 1915. Special Relativity tells us that mass and energy are interchangeable, linked by the speed of light ($E=Mc^2$), again so simple but so profound. General Relativity makes the extraordinary leap of linking time and space into a 4-dimensional continuum and explains gravity as the result of a mass warping the continuum such that other masses move towards it following the path of lowest energy. Very simply, picture a heavy bowling ball resting on a trampoline mat and making a dip in it, then rolling a golf ball in from the side which then spirals down the dip towards the bowling ball and that is gravity in action (kindergarten version).

Albert Einstein is one of my heroes. He used thought experiments to make clear complex concepts ([2]) such that even I comprehend them – the difference between Einstein and myself being that, unlike me, he could go on to express these concepts mathematically and so use them to explore the fundamentals of the universe!

Where does this get us? Well Einstein predicted that the universe must be expanding (to his own acute discomfort because as a religious man he wanted to believe that, as God's creation, the universe should be complete and eternal). Sure enough however, within a decade or so Edwin Hubble, using telescopes at the Mount Wilson Observatory in the U.S.A., and a methodology developed by Henrietta Leavitt (in 1908), showed that the universe is indeed expanding.

It was then realised that if the universe was expanding into the future, then surely one could track that growth backwards and conclude that the universe had a starting point! Wow what an earthquake of an idea – at the very least it contradicts the Biblical story of Genesis!

To cut a long story short there is now near complete consensus that the universe as we know it came into being 13.8 billion (10^9) years ago.

Time for some definitions:

The speed of light (c) = 300 000 Km per second

One Light Year = 60 seconds x 60 minutes x 24 hours x 364.25 days x c = 9.441x1012 or 9.5 trillion kilometres.

A parsec is the largest unit of length we use. It is complex to define but equals about 3.26 light years or 31 trillion kilometres.

2 Consider a man on a train bouncing a ball on the spot. For him the ball is going straight up and down. Someone sitting on a seat on a platform watching the train go past, however sees the ball hit the carriage floor at one point but the next time it hits the floor the train has moved on down the platform – from his point of view the ball is moving in a curving motion in the direction of the train. That is relativity in a nutshell.

So now we approach one end of the scale: the size of the universe. If the age of the universe is 13.8 billion years then the distance from us to the edge of the visible universe must be 13,8 billion light years, right? Well, no, because the universe has been expanding all that time, so the mathematicians agree that the actual distance to the discernible edge of the universe at present is about 46 billion light years from earth in all directions. Thus, our **discernible** universe is a sphere around us of diameter about 96 billion light years, *ie.* 96 x $9.4x10^{12}$= **$9x10^{14}$km**

Think about that for a minute. As a thought experiment if we suddenly moved our earth 46 billion light years away from where we are now, we would probably still be at the centre of a 96 billion light year spherical universe, but the present position of our sun would then be right on the outer edge of that universe!

So, we learn two things: we can never know how big the universe actually is and although we are at the centre of our discernible universe, we humans and our earth are not "special"., any point in the universe will be the centre of its own discernible universe. religions have a problem with this concept. We call our discernible universe the **Cosmos** and we have been exploring **Cosmology.**

Let us stop here, step back and turn in the opposite direction.

How small can we go?

Well, how about "infinitely" small? Sorry, not precise enough! ([3])

Quantum Theory, unlike Relativity, sprang from seeds sown around the end of the 19th Century, but has developed over time and is still developing as I write. It explores the very fundamental structure of the universe and the particles/waves of which it is formed.

For now, we can jump to the work of Max Planck at the end of the 19th Century. A German theoretical physicist working at the Friedrich-Willhelms-Universitat in Berlin, he used Newtonian

3 Zero and infinity are concepts, not integers

physics (Relativity had not yet been developed) to develop a series of units (Planck units) to describe the smallest measurements possible to make, then went on lay the foundations of Quantum Theory.

The connection with Relativity is that we now believe that the so called "Planck length" is the point at which Euclidean geometry breaks down and spacetime becomes what is known as a Quantum Foam ([4]). At present the interface between Einstein's Relativity and Quantum mechanics is Gravity.

The weakest of the four main forces in the universe, gravity alone has (as yet) not been shown to be quantum in character. There is no agreement about the existence of a gravitational field or associated particle. In this observational void two major theories contend for description of the universe at the Planck scale: String Theory and Quantum (Loop) Gravity.

When the mathematics is done (not by me!) the Planck length is calculated to be about **1.6×10^{-35} m**

So, when we consider that the scale of the universe ranges from roughly 1×10^{-35}m to 1×10^{15}m, *Homo sapiens,* at 1m, is 70% of the way along that axis. That is to say that we are 10 trillion times (10^{10}) closer to the size of the universe than we are to that of the fundamental units of spacetime ([5]).

From our place in the scheme of things, the vastness of "outer space" pales into insignificance when compared with the minuteness of "inner space".

4 There is no such thing as a "perfect" vacuum – Quantum Theory predicts a real probability that when no matter exists, energy will spontaneously convert into matter particles which instantly annihilate each other and revert to energy. This process is what is referred to as a Quantum Foam.

5 It is really important to make sensible approximations to huge numbers such as these in order to see the "big picture)

Chapter 2

Time to come "down to Earth".

Currently the Earth is believed to have formed about 4.5 billion years ago at about two thirds the present age of the universe. Life, it is suggested, may have emerged as early as 4 billion years ago. Time to go from Cosmology to Philosophy.

All of us think of ourselves as "alive" and recognise that at some point in time we will cease to be alive (we will die). However, let's look deeper.

There is no absolute definition of Life available to humanity at this time. As far as we know it is a phenomenon associated with planet Earth alone. Abiogenesis is a term now used to describe the emergence of living things from non-living precursors, but the point of transition is not yet defined. Viruses hover in this transitional region. Similarly, death is understood to be the apparent loss of life, but the detail of the process and exact point of onset escape us too.

Having acknowledged that however, much has been learnt about Life.

We know how to recognise it when we see it and we assign it some properties. In 1953 Harold C. Urey and Stanley Miller did an experiment where warm water and a mixture of the gasses: water vapour, methane, ammonia and molecular hydrogen were pulsed with electrical discharges (simulating the primitive ocean, atmosphere and lightening) for a prolonged time.

After a week they found traces of amino acids in the water, confirming that the chemical building blocks of life could have arisen naturally on the earth.

The earliest fossil evidence of living organisms comes from structures known as stromatolites, deposits formed by blue-green algae, dating back 3.5 million years. Stromatolites have been found in the Pilbara, and amazingly, living colonies of stromatolites still exist in Shark Bay, both in Western Australia.

Under the heading of Life/Biophysics in Wikipedia, scientists (including Erwin Schrodinger) are quoted as describing life as *"..a member of the class of phenomena that are open or continuous systems able to increase their internal entropy at the expense of substances or free energy taken in from the environment and subsequently rejected in a degraded form"*.

Why so obscure and complicated? Well as I said above, we cannot define life-the best we can do is describe it and this description gets to the heart of things. Time to introduce another of my heroes: Carlo Rovelli.

If I had to decide between the (literally) thousands of books on the shelves in my study, I would identify two volumes by Carlo Rovelli as amongst the most meaningful for me – "Reality is Not What it Seems" and "The Order of Time".

The first of these books is an elegant and scholarly progression through science from pre-Christian Greece to the present day, in the course of which he makes the case for Quantum Gravity, and in so doing, moves me into this "camp".

The second explores Time, points out that none of the fundamental formulations of relativity and quantum mechanics require time as an input and so shows that time does not exist at a fundamental level in the universe.

Looking at my last sentence I am astounded and mystified by what I have written but, at the coalface of Cosmology, it is generally accepted – Einstein understood it to true.

Rovelli reproduces one single equation in this book, *viz.* $\Delta S \geq 0$ which says that Entropy (S), which represents the degree of disorder in a system, always increases: an intact egg can be scrambled but a scrambled egg can never turn back into an intact one. This is an example of the Arrow of Time where we cannot go back into the past.

What Rovelli proposes here is that the concept of time which humans refer to, observe, measure and interpret in things like past, future, and memory, has emerged as a result of our crevice in the universe initially experiencing a condition of high entropy which provided the precondition for, and powered the evolution of LIFE, leading to the development of us and our development of the concept of time.

It would be presumptuous, pointless and probably impossible for me to undertake an explanation of all that is implied in these last few sentences, so I won't try! Read the book!!!

However, I will say that it leads me to reflect on evolution and come to the realisation that it is the height of hubris to assume that we, *Homo sapiens* are actually the ultimate pinnacle of the evolution of life. Given our monumental ignorance of the perceptive powers of a pantheon of other organisms from octopi to elephants, our claim to be the pinnacle of evolution at this point in time is unsubstianted.

Anyway, I hope that we will continue to evolve, perhaps to *Homo humanus*[6] although the hypocritical lunacy of many of the rulers (chosen or self-imposed) of humanity at present, may result in our branch of evolved life becoming extinct. Regrettably we may even take many other species with us in a violent mass extinction. I hope not, but when you, my Grandchildren, reach my age in 60- or 70-years' time, maybe the future may be clearer.

For now, this perspective causes me to reject all forms of organised religion. I remain open to the idea that some other force may be in existence; how could I not, when we are now aware that all we see in the universe around us constitutes only about 6% of its contents – Dark Matter and Dark Energy seem to make up the remaining 94%, but we do not even begin to understand what they are.

---

6 My mentor for things Latin is my great friend Peter Bryce who is in the grip of Parkinson's Disease and to whom I send my love and best wishes. Without his counsel, *humanus* (meaning kind, humane) is my best guess.

Leaving aside our possible future extinction, let's think about the world as we see it today. It is our astounding good fortune that a man called Charles Darwin was able to rise above the maelstrom of daily life and conjure up an explanation for how the plethora of life forms we see around us came into being. His book "On the Origin of Species" is the foundation on which a vast body of knowledge has been developed regarding the life forms we see around us.

Strangely, the more we look the more it seems that life on Earth was determined to emerge. Intrepid explorers piloting the most highly developed submersible craft at enormous depths have seen, in the lights of their craft, grey plumes (named smokers) emerging from the seabed. These turn out to be thriving colonies of micro-organisms living in pitch darkness and at immense pressures, drawing energy from volcanic sources. These and the organisms found in hot springs, boiling mud pools and caldera of volcanoes make up the Class we call extremeophiles and suggest to me that life will out no matter what.

Chapter 3

Time to have a think about the two "C"s – Conservation and Climate Change.

First off: do these things matter? The answer to that depends on our perspective. In the long run, they do not matter! We know enough about stars to recognise our sun to fall in the class of Red Giants and as such, will eventually run out of nuclear fuel and there will be a vast explosion as the surface of the sun expands out beyond our orbit with the Earth and other planets being subsumed into the sun itself and eventually all that external stuff will dissipate in space and the residue of the sun will contract down to a White Dwarf star. As if that weren't enough, long after that the expansion of the universe (if it continues as we presently observe) will reach the point where everything in space is so far apart that their light can never reach each other and the sky will be a uniform inky black for ever afterwards!

However, before we throw our hands up in despair, the best calculations we have put the time of our sun's demise at least 4 to 5 billion (10^9) years in the future! Since Earth is estimated to be about 4.6 billion years old now, it is only half- way through its existence. Life we believe has been around for about 4 billion of those years but *Homo sapiens* for only about 100 000 (10^5) of those years, and Civilisations for about 10 000 years. So there are vast eons of time ahead for us.

That being so, we should be giving serious thought to conservation, climate change and AI. The best of our Planetary science suggests that Venus is a planet where lack of water in its atmosphere prevents emitted carbon dioxide being sequestered back into its crust, but volcanos keep releasing carbon dioxide into its atmosphere and the rampant Greenhouse Effect has led to surface temperatures on Venus of about 900 degrees (hot enough to melt lead).

On earth volcanos release CO_2 into our atmosphere too, but water dissolves much of this CO_2 and carbon re-enters the Earth's crust as carbonate salts. Trees too, sequester CO_2 from the atmosphere and return it to the Earth's crust to become coal and oil. On a timescale of billions of years this has allowed Earth to retain a climate compatible with life, BUT it cannot respond quickly enough when we humans release vast amounts of stored carbon as CO_2 (by burning fossil fuels *ie.* coal and oil) in the space of a century or two. As a result, the Greenhouse effect will start to change our climate and life on earth will become less tenable. This is inevitable unless we control the level of CO_2 in our atmosphere!

Now, how about conservation?

Again, does it matter?

Again, it depends on perspective!

However, the perspective I adopt is that conservation is the intellectual approach.

Look back at the "timeline of the development of Life" diagram. Life has been around for about 4 billion years (4,000,000,000) we humans for about 100,000 (too short a time to even rate a mention at the end of this spiral). To me it seems obvious that there is a vast amount of information we have yet to discover about the life forms we share the earth with right now. Along with this, many of the environmental problems we are now experiencing source back to our monoculture farming practices.

When we rip up vast swathes of native bushland (which we may reasonably assume to be the result of millions of years of evolutionary selection and so best suited to our land), and replace it with monocultures of wheat etc., problems follow. We divert water from the environment, we introduce vast quantities of fertilizers to maximise production, we flood the environment with pesticides to permit unsuitable plants to grow unimpeded and, as a result, we change our own environment.

Fair enough, we benefit from this by having adequate food for our population, but you know my mantra: *moderation in all things.*

All our native plant, animal and particularly microbial organisms naturally thrived here and remain a priceless reserve of largely unexplored bio-active compounds, so ripping up huge areas of (particularly) tropical rainforests is "throwing the baby out with the bathwater". It is worth remembering that all our first antibiotics were derived by trial and error from natural products from plants or microbes without known use, which today are being eradicated to make room for our monocultures.

In musing on this it seems to me that Civilisations, like fruit, start from very simple beginnings and develop complexity but reach an ideal phase then start to rot. People begin in very rough circumstances, struggling to survive, but gradually the lot of the average person becomes adequate. At the extreme, however, whilst many still starve the most wealthy and powerful become satiated and seek out exotic new experiences. In the James Bond movie, Octopussy, Bond is invited to a meal by the mid- level villain, an aristocratic Indian nabob. The main dish served is individual sheep heads complete with bulging eyeballs and when Bond declines, the nabob grabs an eyeball and pops it in his mouth chewing with gastronomic delight!

Of more relevance here, I understand that in China almost any living thing is food for the connoisseurs including snake meat, insects and live baby mice. Eating a live animal disgusts me anyway, but this quest for the exotic has led to market places where a collection of the most unlikely animals and fungi (sourced often from new environments exposed as new land is cleared) are brought together in "wet markets" in densely populated cities to service these predilections.

It is no accident that the coronavirus now sweeping the world is thought to have made its transfer from its natural host to the human population in just such a marketplace.

Allied to this is the trade in animal products which began as skins providing clothing and features such as elephant tusk ivory being turned into implements and ornaments after normal death of the animal (be it natural or the result of forage hunting). Today, however, these needs rarely exist and the rampant trade in elephant tusks, rhino horn and the skin and organs of the big cats, services the ignorant and an effete minority and threatens extinction of these creatures.

Ignorance, greed, corruption and love of power are ubiquitous characteristics of groups of *Homo sapiens*, like it or not, and I fervently hope we evolve beyond them.

Conclusion

In addition to Quantum Theory and Relativity which describe the way matter and energy interact at the subatomic and everyday levels, a third level of interaction is explored using Thermodynamics.

This is the domain of energy (mainly heat) interacting with bulk quantities of material. Power generators, oil refineries and steam and internal combustion engines are the realm of thermodynamics, manipulated by engineers. This is best discussed with your Grandpa Ettle.

What I would say is that a common thread through all these areas is mathematics. Galileo wrote that the book of the universe is written in the language of mathematics. From the ancient Alchemist pottering over his retort or furnace to the Engineers designing and operating a modern oil refinery the need for measurement and interpretation is ever present.

My advice is to treat mathematics like learning a foreign language and try to find a mentor who can guide you (like going to Germany and learning German by using it daily to survive) rather than, or in addition to lectures. Understanding the point of what you are doing makes learning easier. I wish I had. Artificial Intelligence development is advancing rapidly and technically may outstrip Human Intelligence within your lifetimes, bringing extraordinary benefits but making philosophical concepts of self-awareness and morality crucially important for you.

My point here is that the world is changing, very rapidly. My working world was in the last Century. My parents were "Edwardians" born in the 1900's when King Edward 7th was on the British throne. There were no cars, just horses; candles and gas lighting were the norm and Australia was still strongly bound to Britain.

The 20th Century saw the arrival of steam power, electricity, mass production, cars, aeroplanes, the splitting of the atom, not one, but two World Wars, nuclear energy (and weapons), the Cold War, the Korean War, The Vietnam War, an upsurge in Fundamentalism and a lot of other fighting as well, plastics, the start and end of an era of commercial supersonic air travel, mankind setting foot on the moon, open-heart surgery (like my quadruple bypass a few years ago) and organ transplants, and many other "firsts".

Most of the jobs which employed ordinary people in the 20th Century have either gone or are on their way out. Your cousin Tom Spurling has his finger on the pulse here.

This century (the 21st) is your era.

Good Luck!

References

Any terms I use are described in Wikipedia (but check entries have been peer reviewed). Encyclopaedia Britannica, major university websites and NASA help too.

All books I reference are to be found on the shelves in my study. Below are listed some I have found particularly readable, or just plain fascinating, and all are by well-established authors respected by their peers.

Al-Khalili J and McFadden J (2014), LIFE ON THE EDGE: The Coming of Age of Quantum Biology

Ananthaswamy A (2018), THROUGH TWO DOORS AT ONCE: The Elegant Experiment That Captures the Enigma of Our Quantum Reality

Ball P (2018), BEYOND WEIRD

Barrow J (2011), THE BOOK OF UNIVERSES: Exploring the Limits of the Cosmos

Carrol S (2016), THE BIG PICTURE: On the Origins of Life, Meaning and the Universe Itself

Chown M (2018), INFINITY IN THE PALM OF YOUR HAND

Close F (2011), THE INFINITY PUZZLE: Quantum Field Theory and the Hunt for an Orderly Universe

Cox B and Cohen A (2014), HUMAN UNIVERSE

Cox B and Ince R (2017), HOW TO BUILD A UNIVERSE

Darwin C (2006), ON THE ORIGIN OF SPECIES *(this Folio Society edition follows the text of the First Edition published in 1859 but includes a Glossary of Scientific Terms from the 5th Edition published in 1869)*

Dawkins R (2007), THE BLIND WATCHMAKER *(Originally published in 1986, this Folio Society publication is beautifully presented and illustrated)*

Dawkins R (2008), CLIMBING MOUNT IMPROBABLE *(Originally published in1996, ditto)*

Dawkins R (2009), UNWEAVING THE RAINBOW *(Originally published in 1998, ditto)*

Descartes R (2011), MEDITATIONS AND OTHER WRITINGS *(ThisFolio Society edition is a translation by Desmond M. Clarke with an Introduction by Nicholas Humphrey which I found invaluable as I dipped in and out of the text, based on writings first published in the early 1640s)*

Einstein A (2004), RELATIVITY *(Authorised translation)*

Fayer M (2010), ABSOLUTELY SMALL: How Quantum Theory Explains Our Everyday World

Galileo Galilei (2013), DIALOGUE CONCERNING THE TWO CHIEF WORLD SYSTEMS: Ptolemaic and Copernican *(This Folio Society edition is a translation by Stillman Drake first published in 1953 and copyright to the Regents of the University of California, with an Foreword by Albert Einstein and Introduction by Dava Sobel. The original was published in 1635*

Gott J (2016), THE COSMIC WEB: Mysterious Architecture of the Universe

Halpern P (2012), EDGE OF THE UNIVERSE: A Voyage to the Cosmic Horizon and Beyond

Hawking S (2001), THE UNIVERSE IN A NUTSHELL

Hawking S and Penrose R (1996), THE NATURE OF SPACE AND TIME

Krauss L (2012), A UNIVERSE FROM NOTHING: Why there is Something rather than Nothing

Newton I (2008), PRINCIPIA MATHEMATICA *(THIS Folio Society edition comes with a GUIDE by I. Bernard Cohen, an introduction by Stephen Hawking, a copy of Edmund Halley's "Ode to Newton" and some great illustrations)*

Ostriker J and Mitton S (2013) HEART OF DARKNESS: Unravelling the Mysteries of the Invisible Universe

Penrose R (2011), CYCLES OF TIME: An Extraordinary New View of the Universe

Rovelli C (2017,) REALITY IS NOT WHAT IT SEEMS: THE JOURNEY TO QUANTUM GRAVITY (*Originally published in Italian in 2014 and translated in 2016)*

Rovelli C (2018), THE ORDER OF *TIME (Originally published in Italian and translated in 2018)*

Schoenfeld R (1986), THE CHEMIST'S ENGLISH *(by far the most readable and precise guide to the writing of compact English prose I have ever encountered)*

Schrodinger E (2000), WHAT IS LIFE *(Compiled by the author from lectures given in 1943 at Trinity College, Dublin)*

Smolin L (2013), TIME REBORN: From the Crisis of Physics to the Future of the Universe

Susskind L (2008), THE BLACK HOLE WAR: My Battle with Stephen Hawking to make the World Safe for Quantum Mechanics

Tegmark M (2017), LIFE 3.0: Being Human in the Age of Artificial Intelligence

Watson JD (1968), THE DOUBLE HELIX

Yanofsky N (2013), THE OUTER LIMITS OF REASON: What Science, Mathematics and Logic Cannot Tell Us

Yau S-T and Nadis S (2010), THE SHAPE OF INNER SPACE: String Theory and the Geometry of the Universe's Hidden Dimensions

Zeilinger A (2010), DANCE OF THE PHOTONS: From Einstein to Quantum Teleportation

Postscript

The following are some books on my shelves with personal relevance:

THE RUBAIYAT OF OMAR KHAYAM rendered into English Verse by Edward Fitzgerald with Illustrations by Edmund Dulac; *a lovely old hardback published by Hodder & Stoughton, London with a lot of foxing and that "old book" smell. It's value to me is that it is the only book I have which belonged to my mother!*

LLOYD'S REGISTER OF SHIPPING 1945-46 *has obvious relevance since I was born that year. It is a weighty tome (5.2Kg) and eye-catching because it has a red panel impressed into the cover with:* ***"SECRET see notice inside cover"*** *in gold lettering printed on it. Inside the notice says (in part)* ***"The book and it's supplements are invariably to be kept locked up in the safest place available when not in use. Every effort must be taken to prevent them falling into the hands of Enemy Agents"***

EGYPT AND NUBIA and THE HOLY LAND: Syria, Idumea & Arabia *This two-volume set is part of a Folio Society (2010) Limited Edition (both comprise #675 of 1000 sets). They are huge: 52x35 cm, and heavy: 7.6Kg each, mimicking the size of the original publications by D. Roberts in the 1840's held by the University of Manchester)*

SOUTH POLAR TIMES *A grey upholstered box containing a Folio Society (2012) Limited Edition set (#197 of 1000) of 12 facsimile editions of the South Polar Times. Covering the* *period April 1902 to June 1912, The South Polar Times was a newspaper put together by men overwintering in Antarctica in the first decade of last century, comprising information, humour and art – a great read.*

ANGELS AND DEMONS & THE DaVINCI CODE BY Dan Brown are there *in hardback with beautiful colour illustrations of all art and architecture referred to in the books, along with* THE LOST SYMBOL, ORIGIN and INFERNO.

ANATHEM & SEVEN EVES BY Neal Stephenson *(my favourite science fiction books) are there with seven others by him.*

THE HOBBIT, THE THREE BOOKS OF THE RING SERIES AND THE SILMARILLION by Tolkien *are there in Folio Editions.*

All the **HARRY POTTER** books.

Several titles by **John LeCarre** *including the signed copy of* ***A MOST WANTED MAN*** *I bought when Elise and I went to his lecture in Oxford.*

Lots of other interesting (to me, at least) titles – check them out.

Printed in the United States
By Bookmasters